HOME SPACE CREATIVE DESIGN ROUNDUPS

居家空间创意集 +

古典奢华

CLASSIC LUXURY

深圳市海阅通文化传播有限公司 主编

中国建筑工业出版社

序 PREFACE

At the mention of Luxury and Classics, people will first think of various imperial architecture, which are magnificent and luxury, exquisite and elegant as well as brimming with noble qualities.

The most classical model in European classical structure is the interior design.It stresses interior symmetry and pursues complication and extravagance in particulars. In addition, wallpaper of various patterns is frequently used, such as those with floral patterns and scenes of bible stories painted on. Carpets cover up the wooden floors and painstakingly-embroidered curtains hang from the French windows. Meanwhile, Great attention is paid to the use of adornments including coverlet, veiling and classical decorative pictures on walls.

Luxury and Classics varies in style and form, presenting strong sense of layering. It is elegant and noble, at the same time, solemn and luxurious. Moreover, it is suffused with rich cultural atmosphere and historical accumulations, which makes interior design of classical and luxury style gain attention from an increasing number of elites and people emphasizing life quality.

This book has included many brilliant works of classical and luxury style. You are welcomed to open it and experience the visual impact and lasting appeal of history brought by Luxury and Classics with us.

提到奢华古典，首先让人联想到的便是各种宫廷建筑，它们气势恢弘、壮丽华贵、精雕细琢、庄重优雅，充满了贵族气质。

奢华古典中最经典的要数欧洲古典，室内讲究合理对称，细节繁复奢华，多采用带有图案的壁纸，譬如绘有碎花、圣经故事的场景等；木质地板上多铺有地毯，落地窗上垂落下精心刺绣的窗帘，同时也注重使用床罩、帐幔，墙面上挂古典装饰画等装饰性的物件。

奢华古典的风格形体变化多样，具有强烈的层次感。它典雅高贵，同时深沉而又豪华，具有浓郁的人文气息和历史沉淀，这使得古典奢华风格的室内设计受到越来越多的精英人士或注重生活品质的人所关注。

本书收录了多个精彩华贵的奢华古典设计作品，欢迎您翻开此书，与我们一同感受奢华古典带来的视觉冲击和历史韵味。

居家空间创意集

HOME SPACE CREATIVE DESIGN ROUNDUPS

古典奢华 CLASSIC LUXURY

目录 CONTENTS

THE ANNOTATION OF POSTMODERN LUXURY
后现代的奢华诠释

Design Company: Yi Ge design
设计公司：一格设计

Designer: Liao Xinyao
设计师：廖昕曜

Location: Nanjing
项目地点：南京

Area: 420 m^2
项目面积：420 平方米

The designer grasp the style of villa nicely.The most outstanding character of this house is the decoration which has clear style.Every piece of furniture is exquisite and any of them are the same.All kinds of candlesticks add interesting atmosphere to space.Custom decorative paintings anywhere creat a unique design feature.

From beige wall,soft sofa,carpet to lines of ceiling and crystal chandelier,the whole space is suffused luxury and romantic atmosphere.Living room is spacious and bright,neoclassical furnitures matching vivid decorations and candlesticks,making the space lively,giving more interesting to life.In kitchen and dining room,French windows matching beige ceiling and light color wall which highlight the layering of space.Spacious French windows act in concert with landscape outside the window,creating an atmosphere that harmony with human being and nature.Soft decorations are placed so thoughtfully in the whole building.You can enjoy views one after another at your every pace.

设计师对这套别墅风格的拿捏，体现最显著的是对其特有鲜明设计风格的配饰运用。每一件家具的雕花都精致典雅，但样式绝不雷同；各式各样的蜡台让整个空间增加了几许情调；随处可见的订制的装饰画，则渲染了独有的设计特征。

米黄色墙面、软包沙发、地毯到顶棚上的欧式线条及水晶吊灯，让整个空间弥漫着奢华与浪漫的气氛。整个大厅设计得通透、明亮，别具风格的新古典家具的摆设及精致惟妙的饰品、烛台的搭配，使空间不再单调，使生活更富有情趣。

厨房与餐厅空间独立大块的落地窗结合米黄色顶梁造型和浅色墙面使空间开敞而明亮，极富层次感。宽大的落地窗，使户内户外景观相呼应，打造出现代人追求人与自然和谐共处的生活品位的氛围环境。错落有致的软装将整体建筑空间串联起来，呈现出步移景异的内部意境。

Panasonic

CHANGSHA BIGUI GARDEN EXAMPLE ROOM
长沙碧桂园样板房

Designer：Luo Pengfei
设计师：罗鹏飞

Location：Changsha，Hunan
项目地点：湖南长沙

This model house use marbles,glasses,silver foil and other noble materials,creating a luxury and exquisite atmosphere.

The whole space use white color as its elementary tone with dim lights embellished.Contrasting rose colored bed and concise black wooden table make space spacious and magnificent, like dynamic melody.Besides, irregular grains in bathing room set heavy abstract and modern feeling.

这一套长沙碧桂园样板房采用了石材、玻璃、银箔等晶亮高贵的材料，营造出华贵精致的氛围。

整个空间以白为主色调，其中点缀着昏黄的小灯，玫瑰色床和线条简洁、沉稳的黑色大木桌对比强烈，整体显得非常宽敞、大气，旋律动感跳跃，而浴室中不规整的纹理，显示出浓厚抽象而极具现代感的意味。

VISUAL FEAST

品鉴·视觉盛宴

Design Company：Fame Deco
设计公司： 智上名筑

Designer：Han Zaidong
设计师： 韩在东

Area：749 m^2
项目面积： 749 平方米

This design aims at a businessman who has found some success. Social activity is an important part of his daily life.So,I blend his background and hibit into my design. Rococo defines the style,with understated luxury idea.Graceful lines completely set off the atmosphere. A transparent partition with beautiful shape between living room and hallway gives a hazy connected visual. The layout with various levels sulfuses space with rhythm.

本案设计面对的业主是商业人士，其平时的社交活动是生活的重要一部分，而且是个小有成就的成功人士，所以，设计理念融入了业主的生活背景和生活习惯。首先在风格上的定格是洛可可，设计理念是低调奢华，柔美的曲线把整个设计烘托到了极致。客厅和门厅中间巧妙地利用了一个造型曲线优美而且通透的隔断，让门厅和客厅出现了似断非断的视觉效果，高低不同的错层设计，让空间充满了节奏与韵律感。

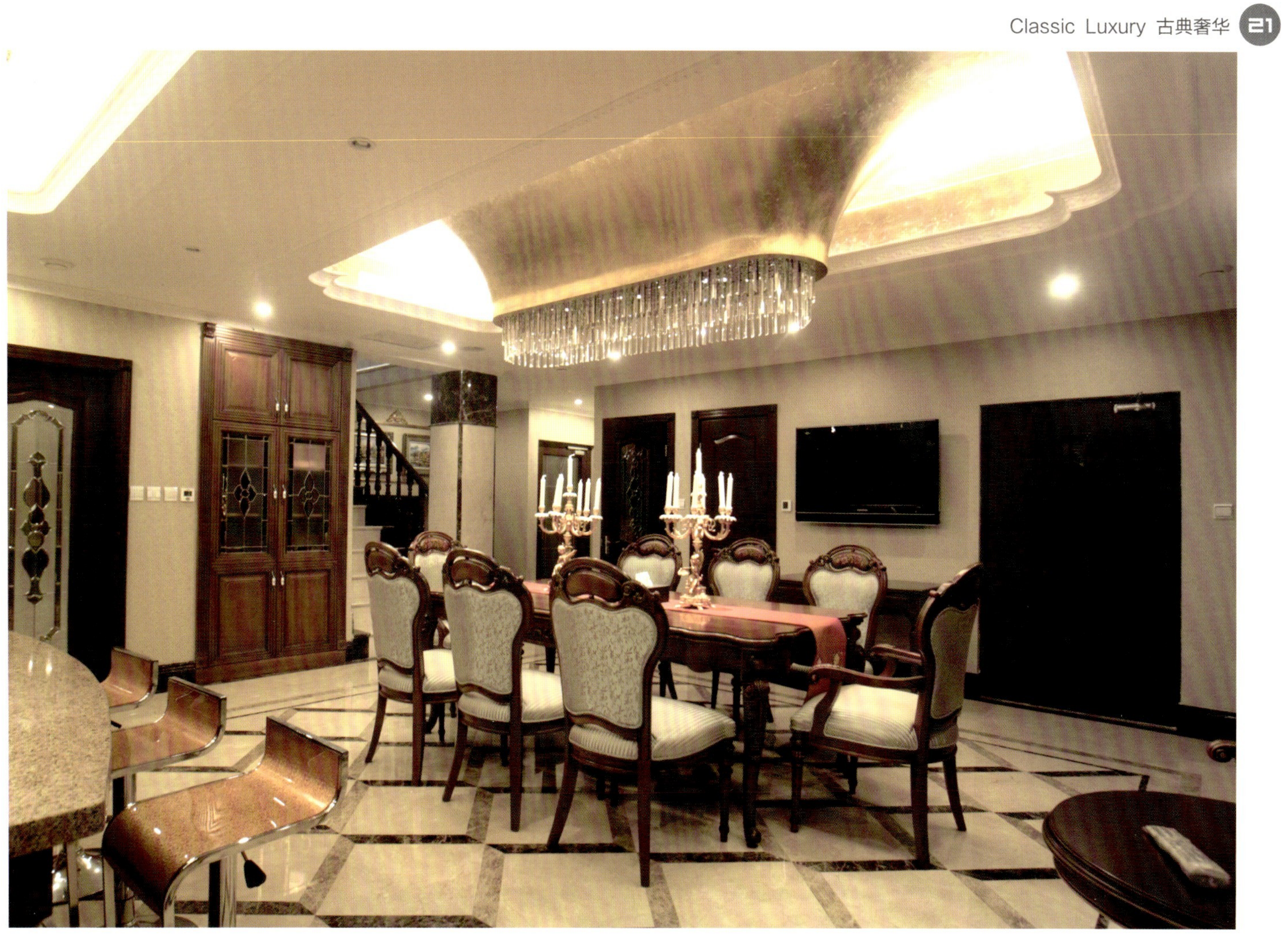

YIYUN EDGE WATER FOUR PERIOD NO.21 FLOOR VILLA
依云水岸四期 21 号楼别墅

Design Company: Guangzhou Ping Ge Interior Design Consulting Co., Ltd **Designer:** Zhou Wensheng **Location:** Foshan, Guangdong **Area:** 667 m²

设计公司： 广州市榀格室内设计顾问有限公司 **设计师：** 周文胜 **项目地点：** 广东佛山 **项目面积:** 667 平方米

European Neoclassical is main style of this villa.The villa is a lengthwise townhouse,it has an independent garden.

Building is only part of the whole design.Interior layout,landscape,building and all rhe other elements in interior are taken into consideration,are divided and refactored in different forms.They are in perfect harmony with European Neoclassical Style.

Golden century marble is the main facing of the wall in public space,and other personalized marbles,like Tiger Onyx, Luise White which have clear,artistic grains are used in parts and add pleasant surprise to space.The room facing mainly use natural ebony, matching with import wallpaper,leather and textile.Hard materials complement soft materials perfectly,and creating an unique magnificent,stately and elegant European Style.

本案的室内空间装饰及装修风格，设计师主要以欧式新古典风格为主要表现手法。整个建筑为一幢四层的并联别墅，建筑外形纵向走向，拥有独立园林。

在整个概念中，建筑仅仅是一部分。室内布局、景观、建筑体以及室内的元素被统一考虑，进行不同形式的切分和重构，并在欧式新古典风格下，形成统一的多重变奏。

公共空间主要以金世纪米黄石材为墙身饰面，局部使用一些来自欧洲的个性石材，如黑海玉、路易斯白，具有清晰、艺术化的纹理纹路，为空间增添不少惊喜。房间主要采用天然黑檀木饰面，配有进口的壁纸、皮革及布艺，硬质材料同软质材料相互搭配，取得了和谐统一的效果，营造出欧式特有的磅礴、厚重、优雅与大气。

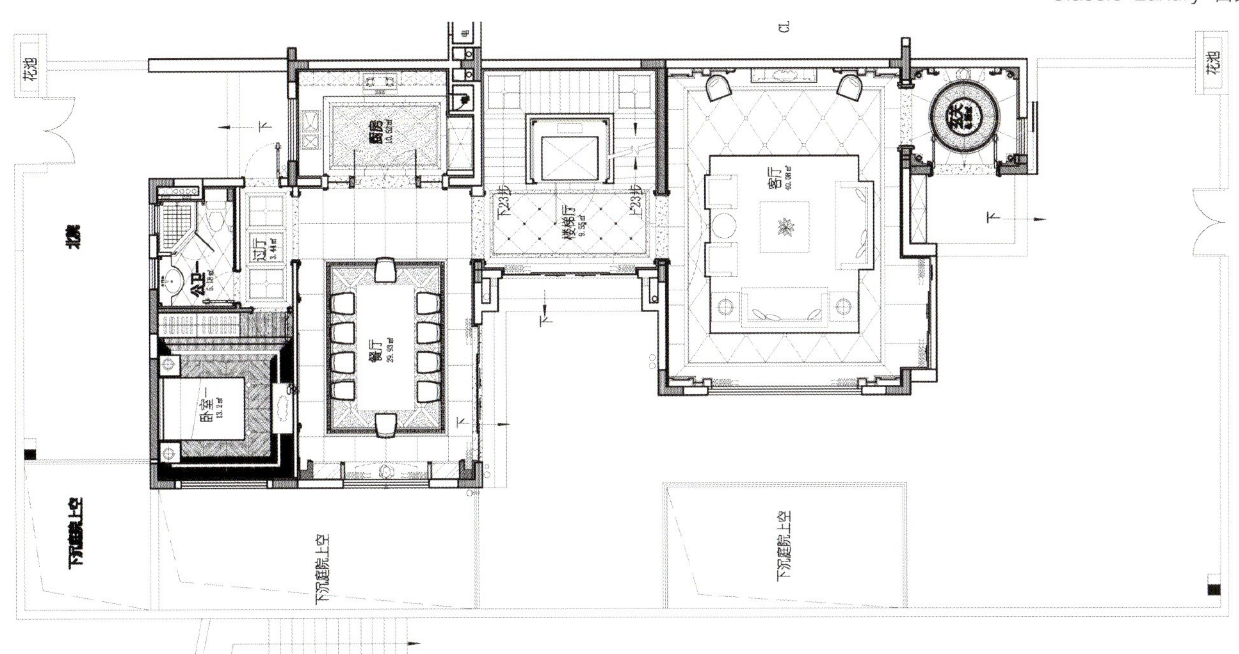

The first floor plan 一楼平面图

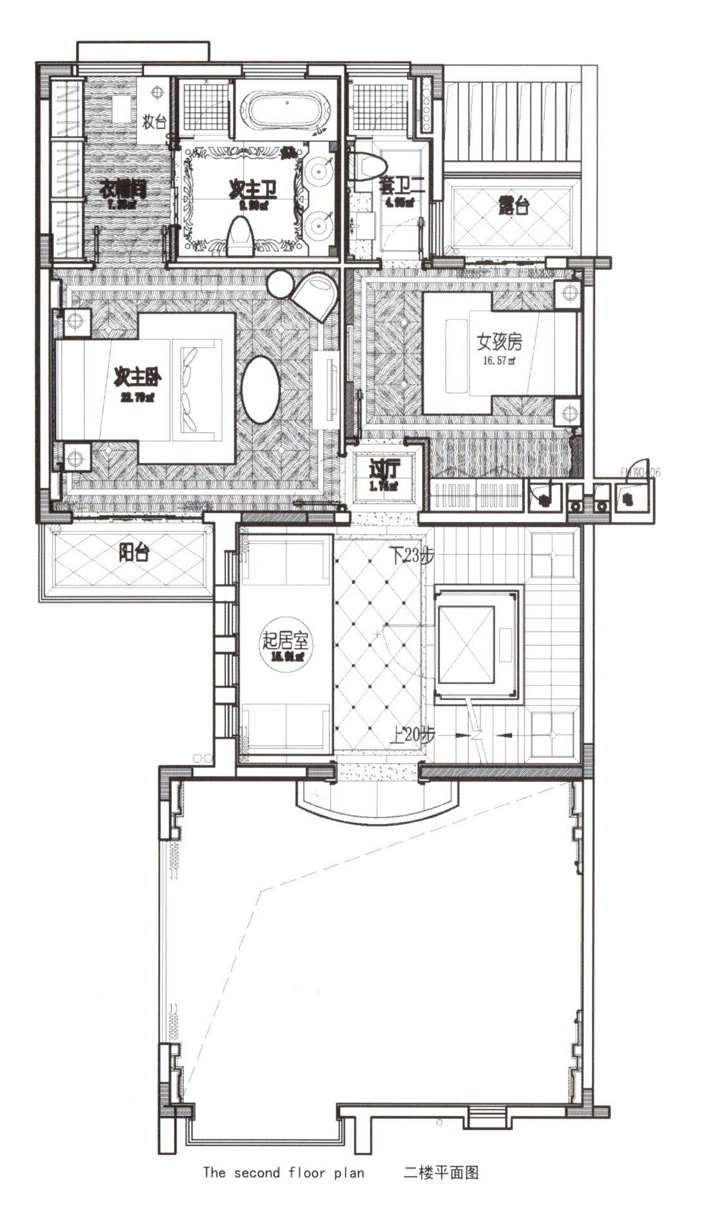

The second floor plan 二楼平面图

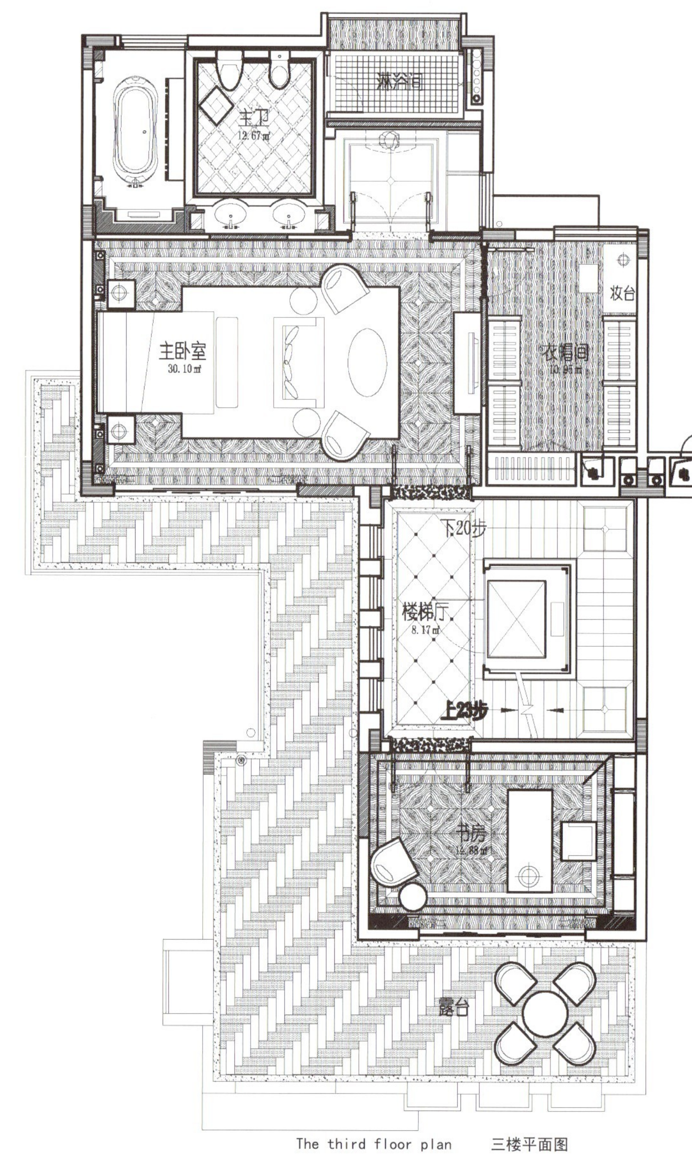

The third floor plan 三楼平面图

DA VINCI

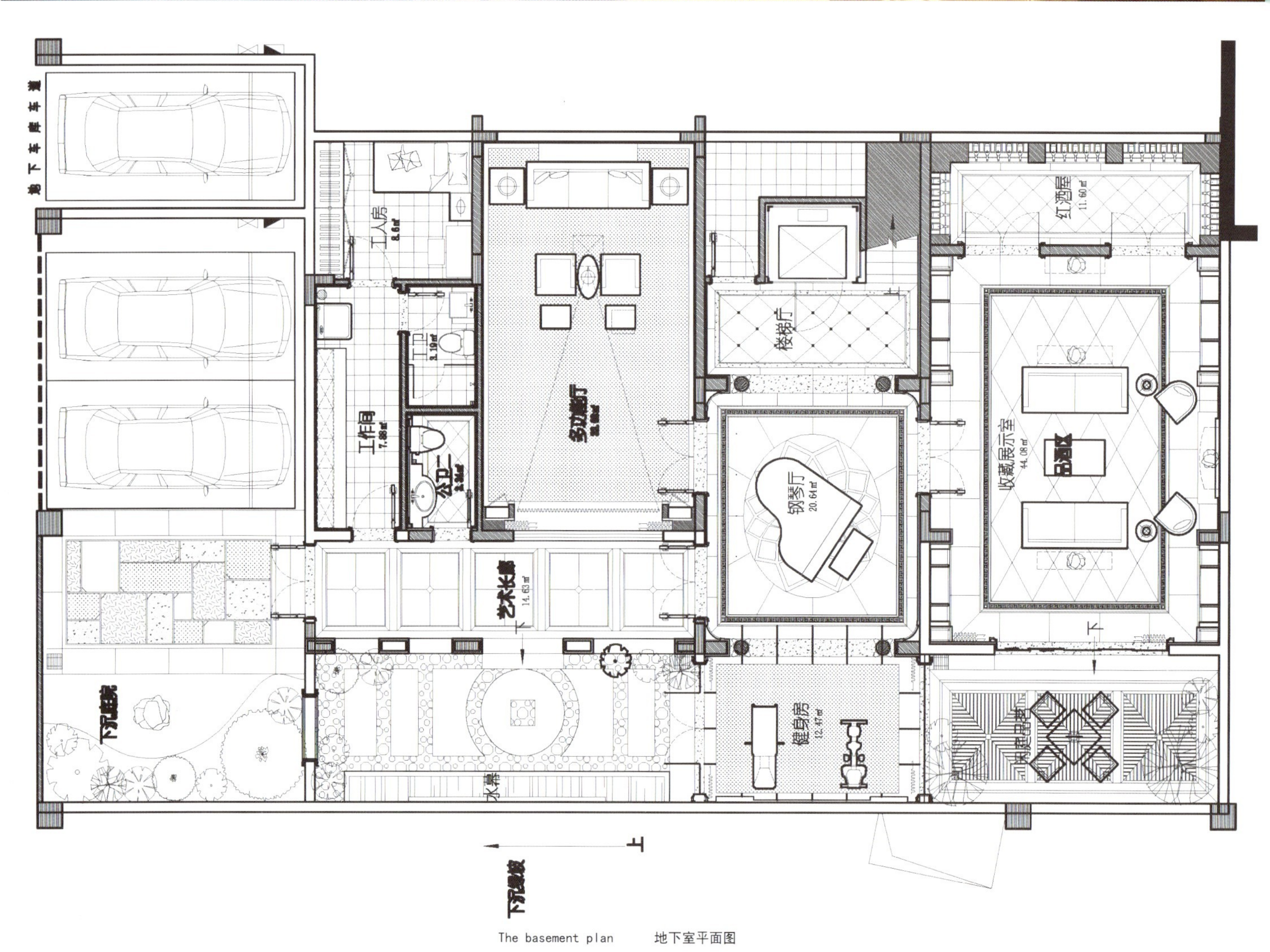

The basement plan　　地下室平面图

QIANDAO YACHT VILLA D, HANGZHOU
杭州千岛湖墅游艇大宅 D 型

Design Company: Ivan C. Design Limited
设计公司： Ivan C. Design Limited

Designer: Ivan Cheng
设计师： 郑仕樑

Location: Hangzhou
项目地点： 杭州

Area: 800 m^2
项目面积： 800 平方米

This design mix "luxury" and "beauty" subtly.Luxury decorations and exquisite wall set a classical flavor,showing the owners dignified status. Noble atmosphere complement modern home life perfectly.Modern methods keep the meaning and discard the forms,creating this neoclassical style.Furnitures show to be stretching and graceful,like free flowing water, deducing a stately and elegant style.

In this elegant color tone,sofa and decoration adopt other colors,having a distinctive feeling.The whole space suffuses a slight exotic flavour,being tranquil and philosophic.

整个设计将“奢”与“美”巧妙结合，奢华的装饰和做工细致精美的墙壁很好地衬托出复古的古典气质，彰显业主至高无上的尊贵地位，整体高贵的气质空间结合现代的居家情怀，运用现代的表现手法，存意舍行，新古典风油然而生。家具线条舒展优美，简洁流畅，犹如行云流水，演绎了庄重而优雅的风格。

在高雅的色调中，沙发和装饰运用了其他彩色，仿佛藏着无数的禅机，总是温馨、总是跳跃，别有一番风情。空间散发着淡淡的异域气息，同时禅味十足，静谧而哲理独蕴。

QIANDAO VILLA F

千岛湖墅 F

Design Company: Ivan C. Design Limited
设计公司： Ivan C. Design Limited

Designer: Ivan Cheng
设计师： 郑仕樑

Location: Hangzhou
项目地点： 杭州

Area: 836.71 m^2
项目面积： 836.71 平方米

This villa locates in Qiandao lake which called “the Most Beautiful Lake under Heaven”,adjoining the Xianwan international resort.It is one of the 26 palcace standard villas.Qiandao villa has great momentum,blending with surroundings.

The villa adopts neoclassical style,having exquisite decorations,setting free and magnificent features. The areas is 836,71 sq.m. AR-DUE cabinet,Italy FIR tap, Turkey VitrA sanitary appliance and bathtub,leather and silk cloth which are classic brabds and are all full of history,setting owner's state and wealth.Design mix Luxury and Beauty together perfectly,luxury decorative materials and beautiful art furnishings,making up space reasonably,bringing nice life to host.

千岛湖墅 F 择址于“天下第一秀水”美誉的千岛湖畔，进贤湾国际度假区黄金一域，为稀世 26 席宫殿级别墅之一。千岛湖墅封鼎千岛，自成“千岛墅王”气韵，与周围湖景山色浑然一体。

千岛湖墅 F 设计采用新古典风格，装饰精美，体现宫廷俊朗而庄严的特征。设计面积达 836.71 平方米。在设计选材上，选择拥有悠久历史、享誉世界的纯正经典品牌艾度维橱柜、意大利 FIR 水龙头、土耳其 VitrA 洁具和浴缸，真皮、高级丝绒布匹等，彰显主人不凡的财富和地位。设计中将“贵”与“美”巧妙地结合，奢华的内饰材料，优美的艺术陈设，合理点缀空间，带给主人“美”的生活享受。

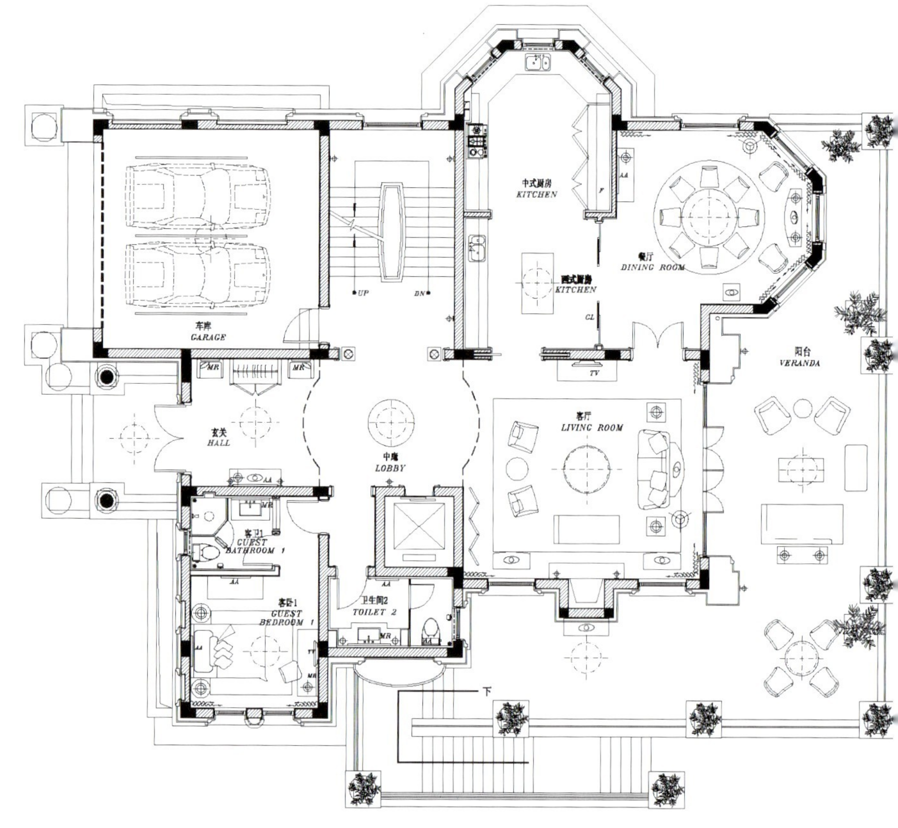

The first floor plan 一楼平面图

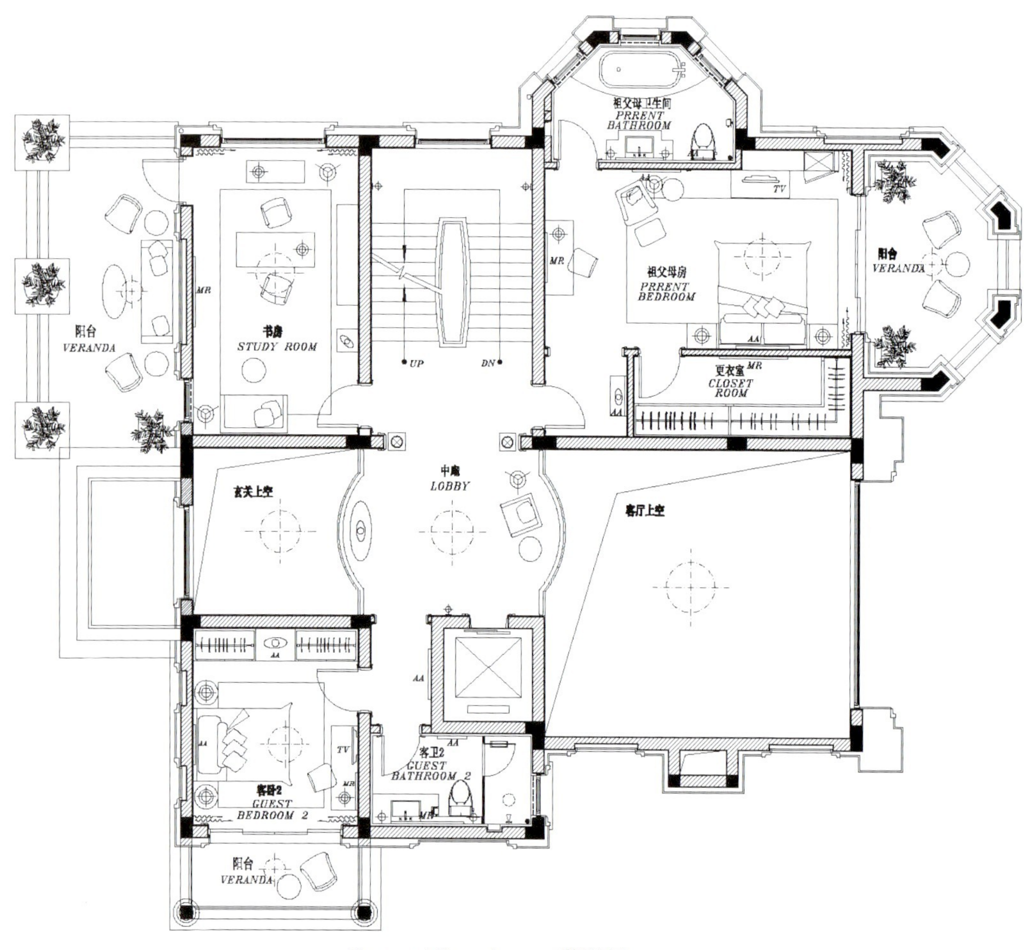

The second floor plan 二楼平面图

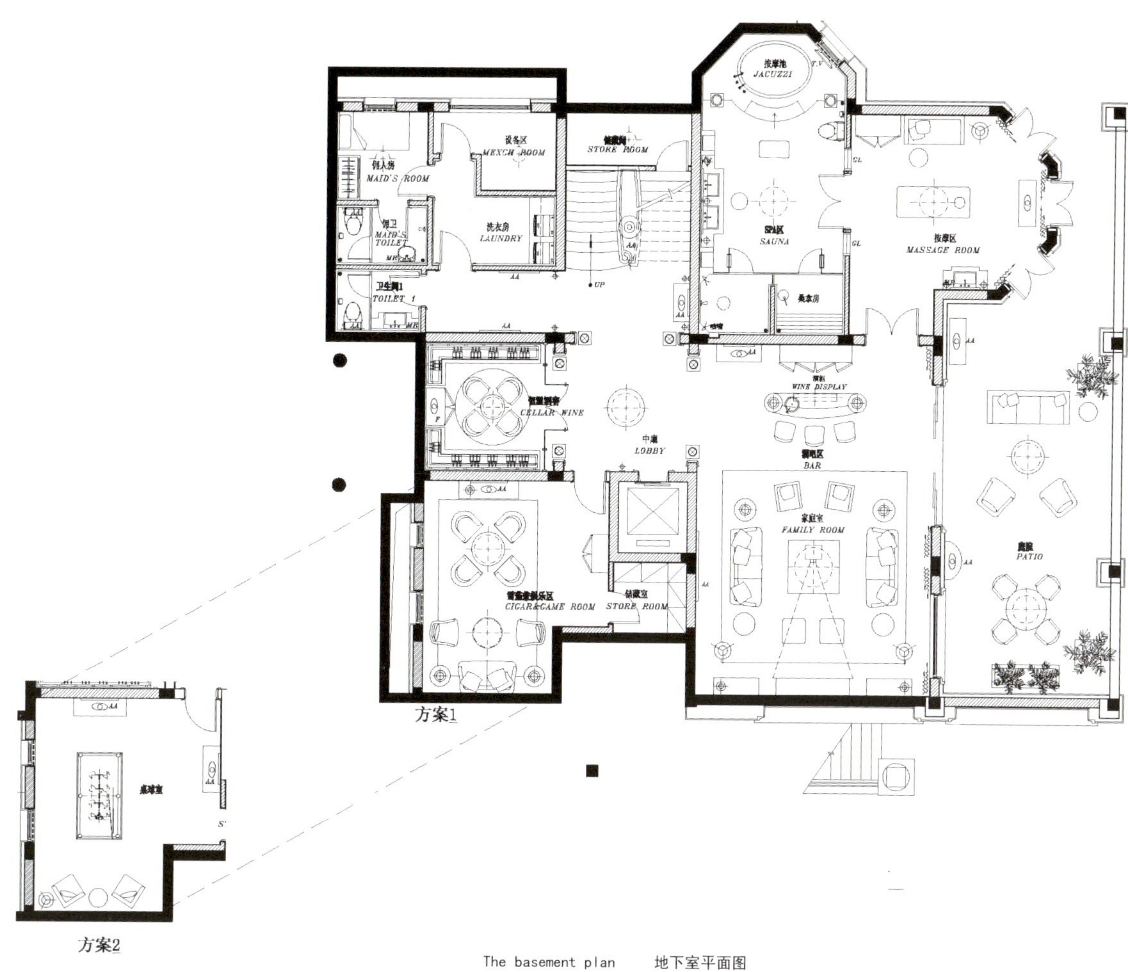

The basement plan 地下室平面图

SI JING YI JING YUAN

泗泾颐景园

Designer: Zhou Chuang
设计师：周闯

Location: Shongjiang
项目地点：松江

Area: 260 m²
项目面积：260 平方米

Luxury ,privateness, gorgeousness,personality are the definition of a villa. This design is European style.Spacious living area, space layout is reasonable in generatrix.We make change in stairs,using the modern way to divide the hallway,living room and dining room,to make sure that we can get sunlight and air easily,the landscape outside can blend into interior.The design method is concise with luxury materials.

对于别墅，尊贵、独有、华丽、个性是在定义上。欧式是我们赋予本案的设计主题。较为开敞的生活区域，注重动线及各空间之合理安排，在楼梯的位置上作了变动，将门厅、客厅、餐厅利用现代的设计手法加以分隔，使空间得到完整的划分并注意采光及通风，将外部景观融入室内环境。本方案设计手法比较简洁， 不过在材质上是华丽的处理。

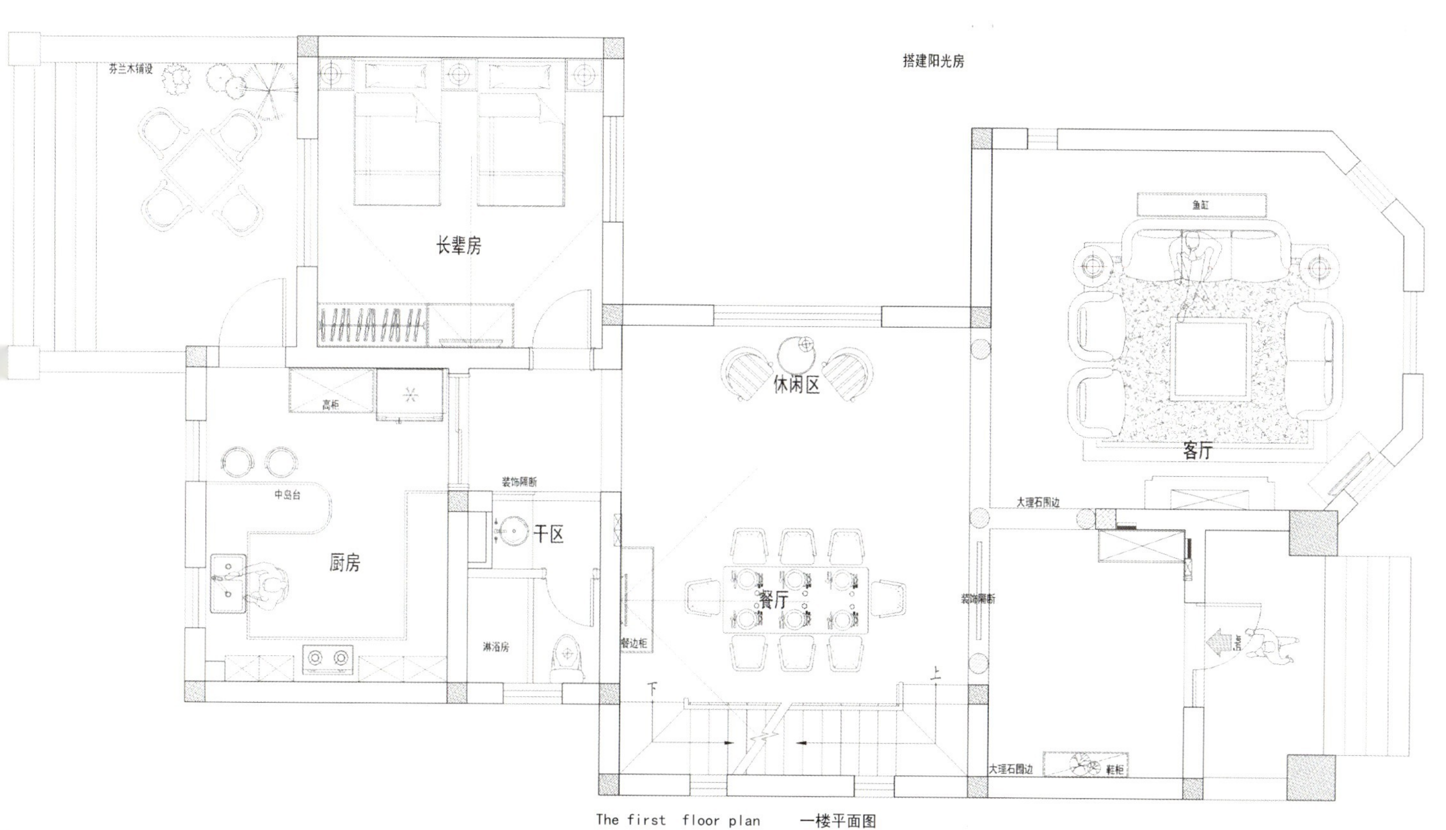

The first floor plan 一楼平面图

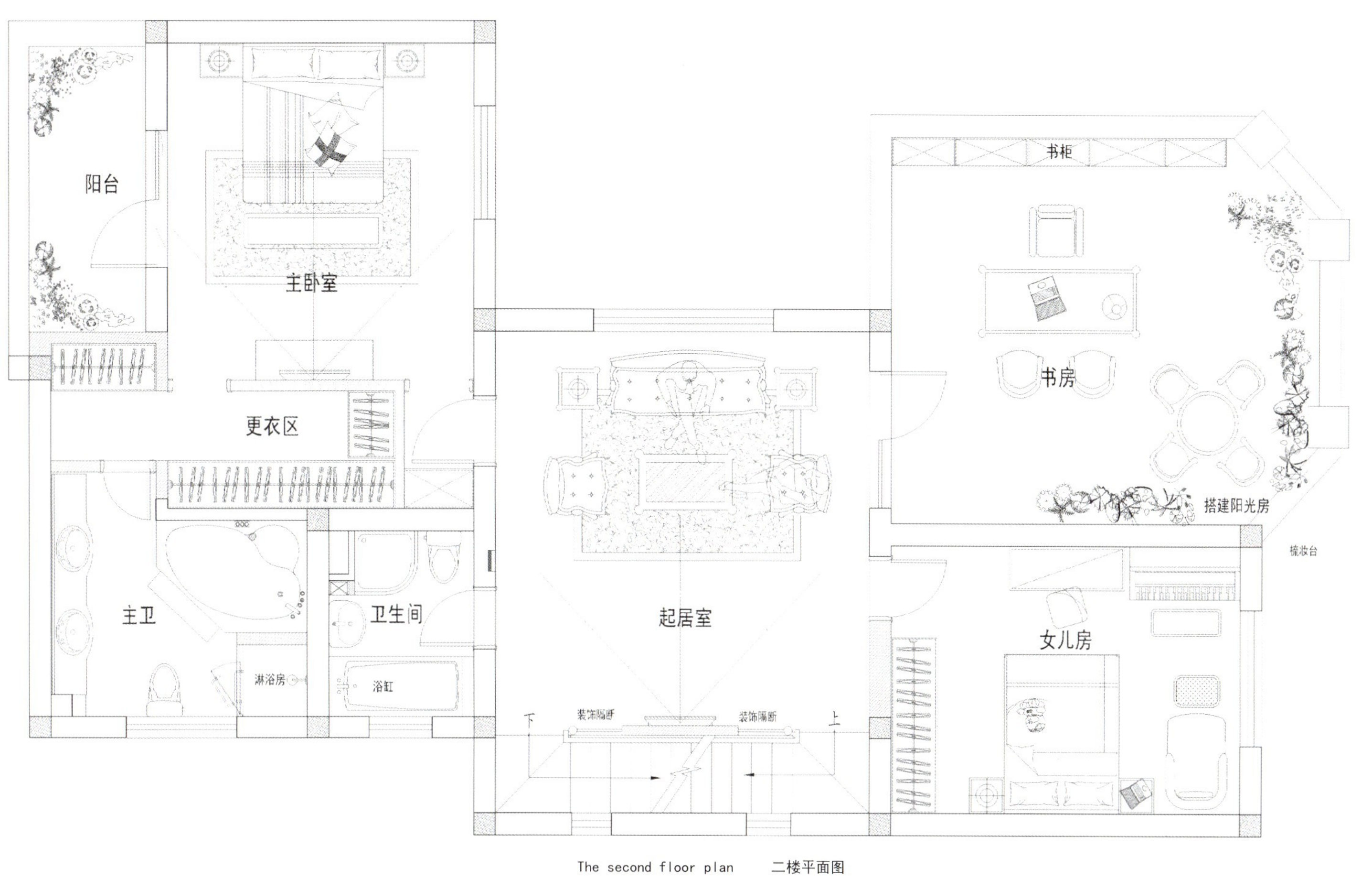

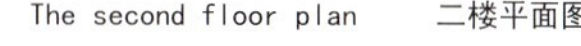
The second floor plan　二楼平面图

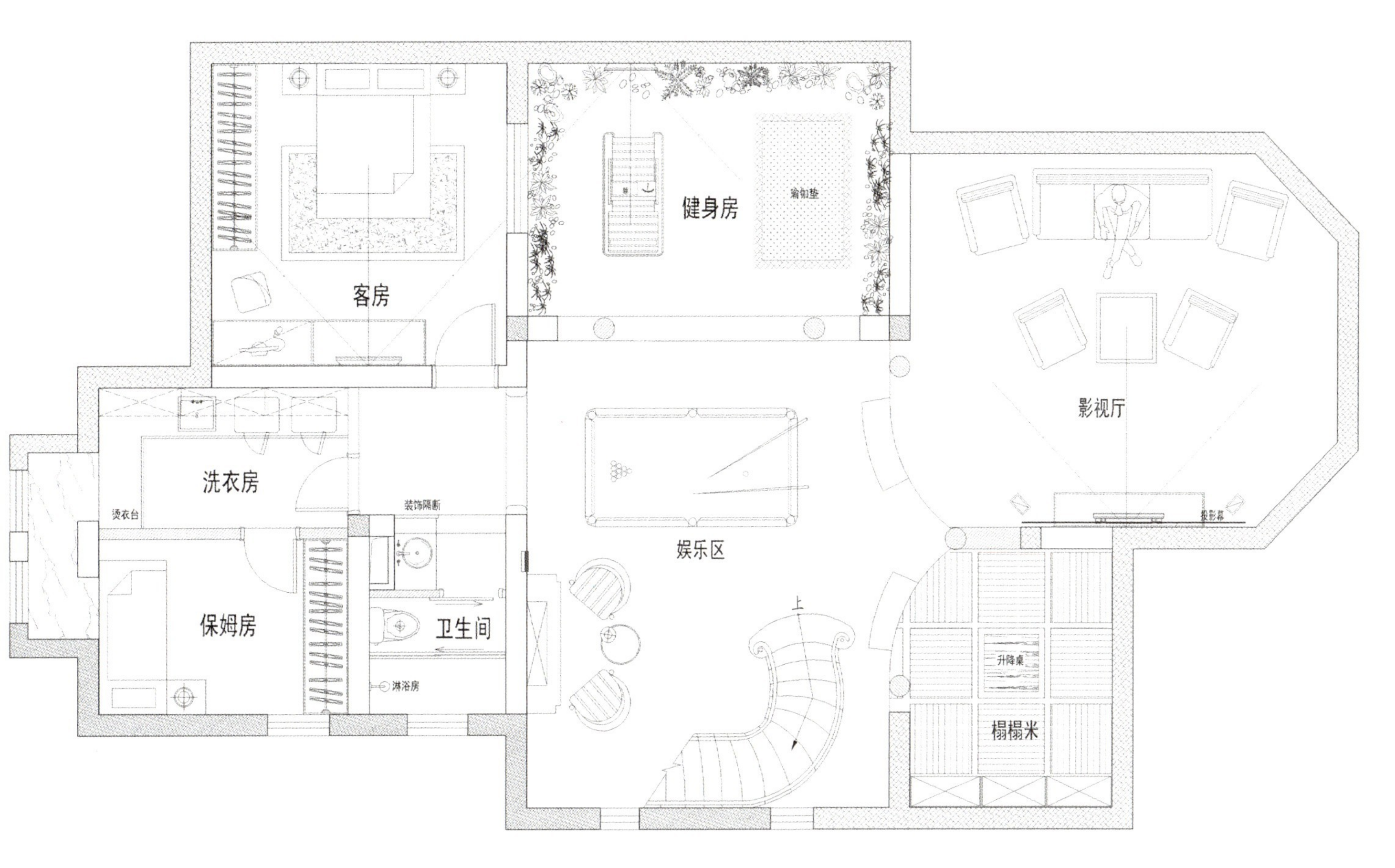

The basement plan 地下室平面图

WAN DA
万达

Designer: Dai Wei
设计师： 戴伟

Area: 90 m^2
项目面积： 90 平方米

The most important is a cozy life no matter the house is big or small. Designer creates not only a room to live,but a luxury and happy paradise.This design aims to build an elegant culture space in premise of original structure.Considering owner's status,designer uses modern and modest way to interpret classical luxury style,meanwhile,suffusing elegant lift taste.

Besides stone and wood,metal and glass and other modern materials are applied to improve contrast effection.Large soft leather present luxury and modern feeling,tranditional dull atmosphere is broken with warm light.Layering sense and penetration of space are enlarged.

Soft decorations including lights and furnitures are sophisticated,expressing modern sense with western elements. Crystal chandelier match European classical furnitures,round mirror,haning painting and decorations perfectly and subtly,setting luxury European style off completely.

房子不在大小，重在营造出居住空间的舒适氛围和居住情怀。设计师不仅为居住者打造了一个生活需求空间，更为他建造了一个奢华居室、让人伸手可及幸福的空间。本案旨在尽量保持原户型结构的同时，营造一个蕴含文化气息的雅致空间。考虑客户的定位需要，整体风格是用现代而稳重的手法演绎古典的奢华主义，同时又不失高尚的生活品味。

在选材方面，除了石材和木材，还用了金属和玻璃等现代材料，大大提升了原质感的对比效果。软包皮革的大面积使用，在表现尊贵的同时还增添几分现代感，配搭暖色的灯光效果，打破了传统沉闷的格调。空间的层次感、穿透性大大增强。

后期软装的配搭，灯具、家具等方面也很考究，既有西式元素，又具有强烈的时代感。顶棚水晶灯的闪烁效果，配搭欧式古典家具，黑白水晶灯、圆镜、挂画、饰品，一切将奢华的欧式风情演绎得淋漓尽致。

NEW CLASSICAL VICTORIA'S SWEET MEET
新古典维多利亚的甜蜜相遇

Design Company: Chengyi Design Alliance
设计公司： 程翊设计 联盟

Location: Taipei
项目地点： 台北市

Area: 165m^2
项目面积： 165 平方米

Totem patterns in hallway cheer up guests' mood.Bright glass materials make sure clear visual thought in conner.Crystal chandliers are hanging in round ceilings.There are defferent way to living room and dining room.

Symmetrical layout,arrangements of living room entrance take full use of corridor,making dining environment more smooth.The transition of private space beyond romantic imagination.The pop wall papers in children's room reflect girls' younth and liveliness.The carved screen hides the door,mysterious and elegant purple color is the most outstanding feature.Wall papers match line boards to fix the position of TV wall.The designer get profound appreciation again with exquisite design and sincerity.

甫入门厅，精品语汇以图腾呈现，转换了来客到访心境，玻璃材质的流动通透，即便是转折中的封闭，也保有了明亮视觉的清新感受。水晶灯在圆形上潜段落悬吊。导引路线也分野着客、餐厅区块。

整齐对称，起居室的入口安排充分利用了走廊，使得就餐环境更为温馨。私人空间转换跳脱浪漫想象，婴儿房内的时尚墙纸表现了小姑娘的青春与活力。雕花板为屏挡住了大门，紫色系的神秘高雅更为突显。壁纸混搭再以线板为框，巧妙定位出电视主墙面，坚持做工精细的程翊设计用满心的诚意，再次获得来访者的由衷激赏。

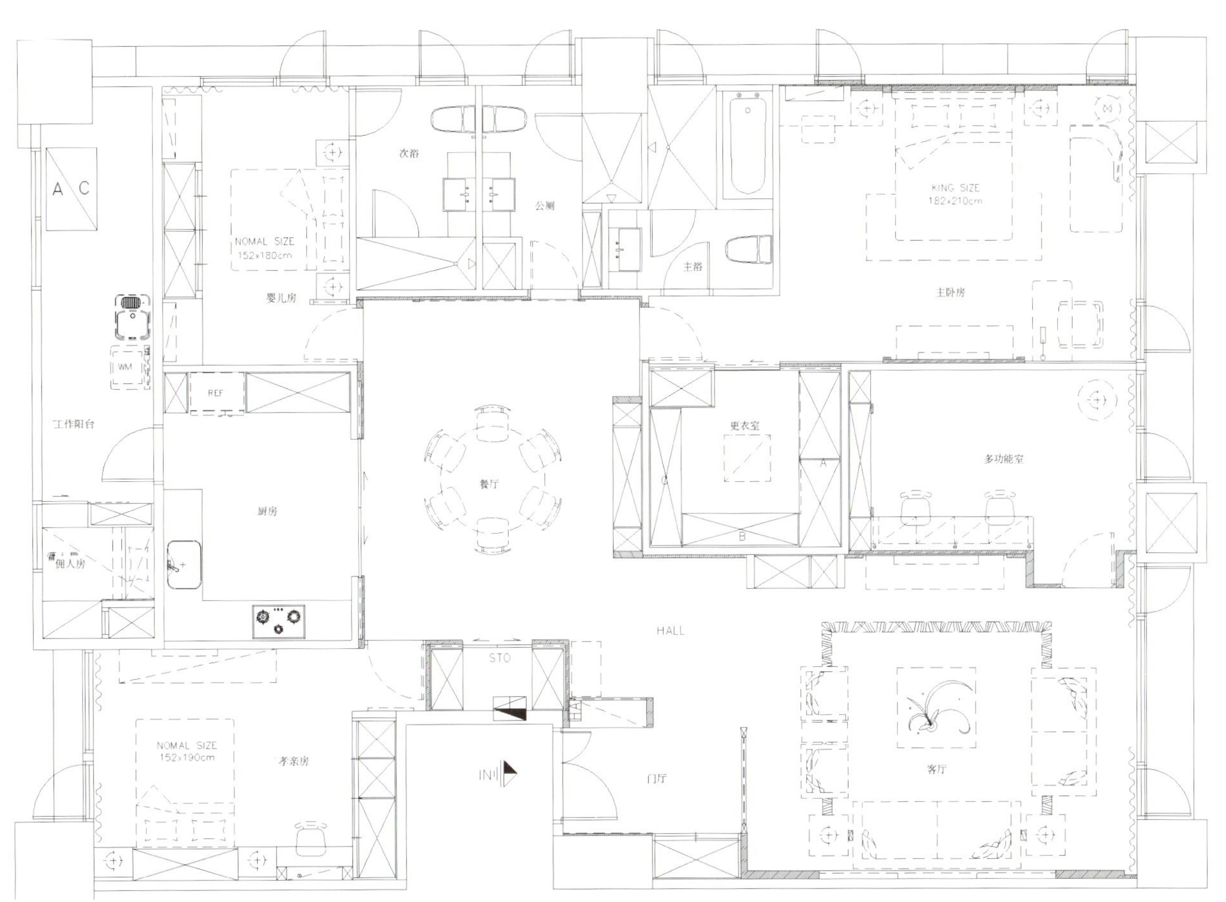
A C
次浴
公厕
主浴
NOMAL SIZE
152x180cm
婴儿房
KING SIZE
182x210cm
主卧房
REF
工作阳台
更衣室
多功能室
餐厅
厨房
佣人房
HALL
STO
NOMAL SIZE
152x190cm
孝亲房
IN
门厅
客厅

Plan　　平面图

EXPERIENCE LUXURY, XUJIA

体会奢华 · 徐家

Designer: Liao Xinyao
设计师：廖昕曜

Location: Jiangshu
项目地点：江苏

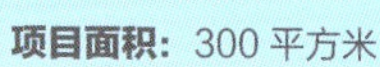

Area: 300 m^2
项目面积: 300 平方米

Living room follows European style.Sophisticated architectural elements add aesthetic sense to space,bright design demonstrates magnificence. Clear outline forms a harmonious picture with splendid colors,exqusite beige wall is fashion and stately.The whole space is not full of gold and silver,but a low key luxury and cozy home.Elegant taste matches luxury life and get a perfect balance.

Every detail likes a flower stretches slowly in the spring breeze.It suffuses fresh smell,and being familiar and gentle.Colors are mixed together to bring a harmonious painting,delivering a warm feeling,this is home.

徐宅客厅的设计风格为欧式，以讲究的建筑元素增加空间的美学，鲜明的线条设计彰显大气态度，清晰的轮廓与华美的色彩相搭配，细致的米黄色墙面时尚又气派，整个空间没有披金带银的奢华，却有不张扬的舒适，使高尚品位与奢华生活达至平衡点。

每一处的细节，如同绽放在春风中的花朵，缓缓慢慢地舒展，每一处都透露新鲜，且每一处都是那么的熟悉温和，融和的颜色块放在一起，就是一幅和谐之美的画面，让人温暖。家就应该是这样。

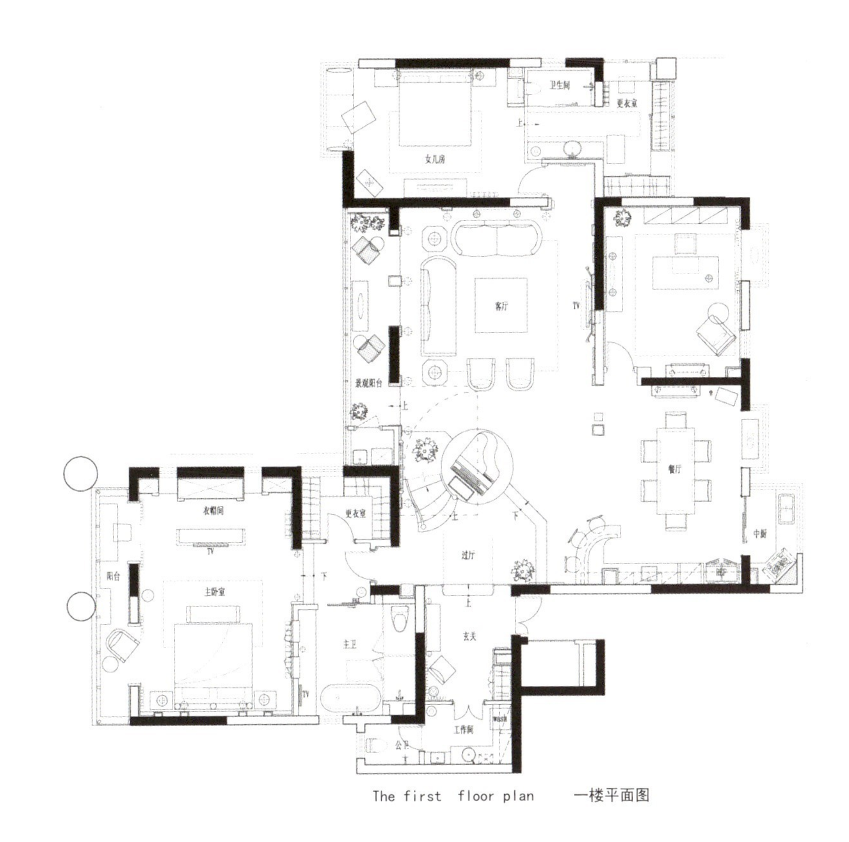

The first floor plan 一楼平面图

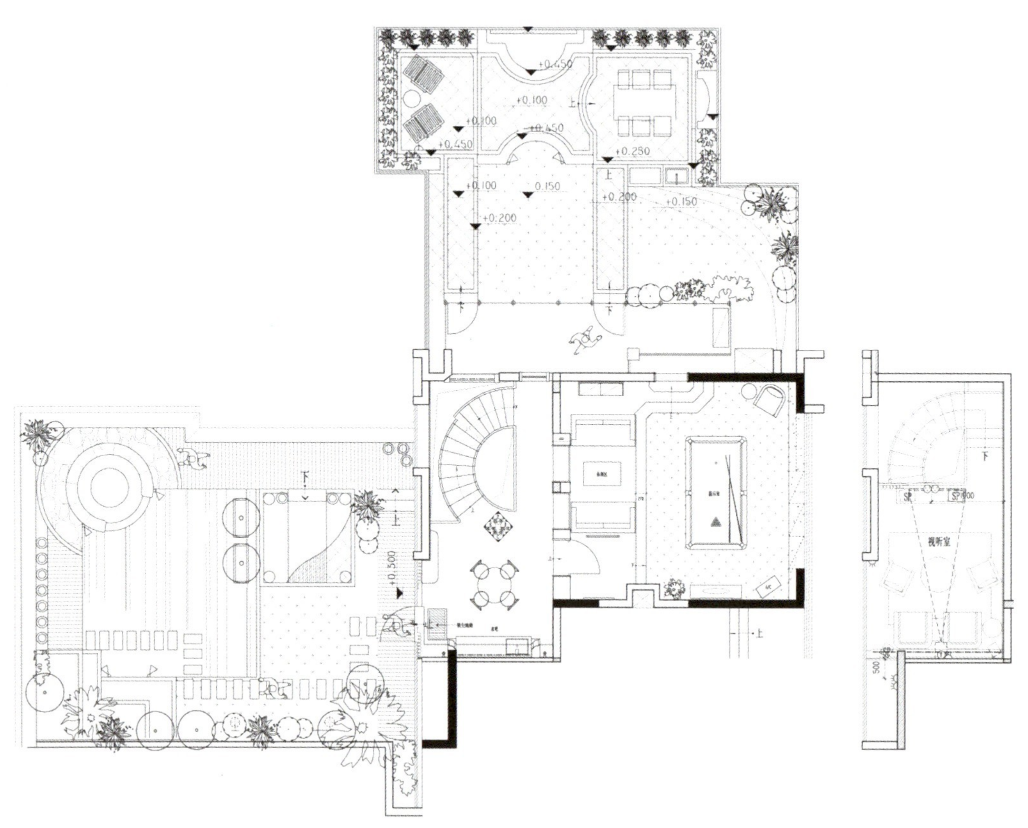

The second / third layer plan　　二／三层平面图

THE MILKY WAY BEND

银河湾

Designer: Zhu Ziquan
设计师：朱自权

Location: Nanjing
项目地点：南京

The best scenery is when flowers are all in blooming.We find an elegant area in the air which is full of the fragrance of flowers.We can feel dignity in dark,touching luxury in elegance,the most is being cozy and romantic. The unique design bring magnificent,vivid,enthusiastic,gorgeous and elegant features to our lives.The house take full advantage of space to divide three bedrooms and two living rooms.There is no storeroom,so designer adds cabinet as many as possible.Living room forms hexagon,which matches the original ceiling,also showing a harmonious beauty.Large scale mirror make space more luxury.Dark pattern wall paper,gesso horn line,European table and exquisite wool carpet are all necessary elements to European style.They push the elegant European style to a new perfection in the united mood.

繁花盛开处正是风景最好时，在这满屋的花香中，我们看到繁花留恋之处正是有一室的清雅在此驻足。在欧式风情的映照下，在这个居室中，深沉里显露尊贵、典雅中沁透豪华，让我们感受到更多的是惬意与浪漫。匠心独具的设计师将宏伟、生动、热情奔放和纤巧华丽、雍容优雅带进人们的生活。这是一个三房两厅的房型，在整体设计上，充分利用了空间。由于房子本身没有储藏间，所以设计师尽可能地增添了储藏柜。整个客厅呈六边形，在顶面设计上不仅要配合原顶面，也要展现出一种和谐与美观。大面积的镜面使整个空间看起来更加奢华。暗花纹壁纸、石膏线脚、欧式桌椅和精致羊毛地毯都是欧式风格不可或缺的元素，在统一的整体基调下，典雅的欧式风格被推向新极致。

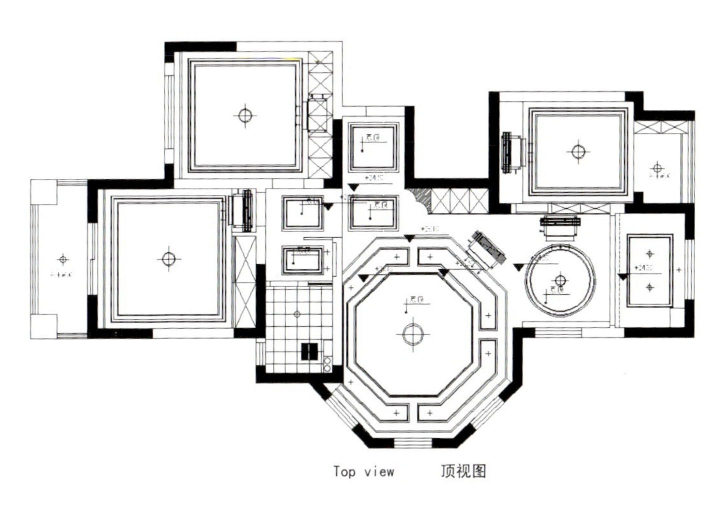

Top view 顶视图

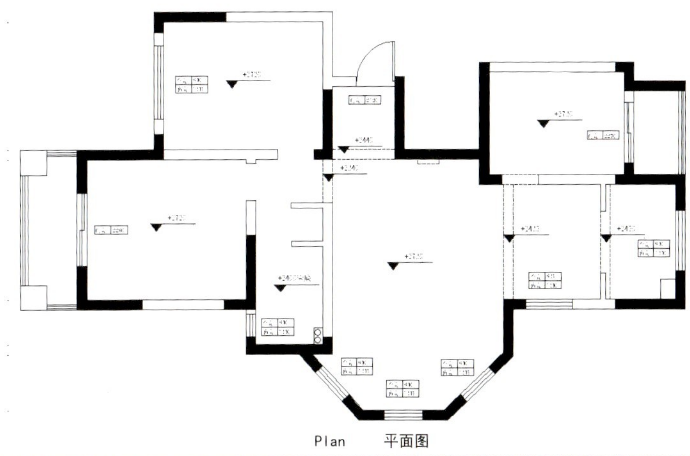

Plan 平面图

ZHONGHAI JADE LAKE SHORE VILLA
中海翡翠

Design Company: Shanghai Big Jue Construction Engineering Design Co., Ltd
设计公司： 海大珏建筑工程设计有限公司

Location: Shanghai
项目地点： 上海

Area: 450 m^2
项目面积： 450 平方米

According to modern European style,this house expresses luxury and elegance completely.The whole house creats a romantic idyllic style without complicated carvings,cumbersome decorations, and heavy elements.Public space use import special painting which has special textures instead of stones and wood. The painting shows a noble,classical and warm atmosphere in light and creats a classic French idyllic style with the main tone.

秉承现代欧式设计风格，将奢华与典雅的美感发挥到淋漓尽致。没有繁复的雕饰，没有繁琐的装饰，没有厚重的欧式元素，整体营造出一种欧式田园的浪漫气息。公共空间的主要建材摒弃了常规的石材或实木，使用了进口的特殊涂料，这种带有特殊纹理的涂料在灯光的渲染下带出了高贵古典、温馨的气息，结合设计的主基调，组合成为一个经典的法式田园风格。

SANXIA –ZHONG YUE

三峡 – 中悦

Design Company: Athens Design Engineering Co., LTD **Designer**: Huang Tingzhi **Location**: Taiwan **Area**: 234 m^2 **Photography**: Lin Fuming

设计公司： 雅典设计工程有限公司 **设计师：** 黄庭芝 **项目地点：** 台湾 **项目面积：** 234 平方米 **摄影师：** 林福明

The white space in hallway sets golden luxury off,thus French classical noble is begin.The house consists of mostly the luxury European Classical Palace Style.Complicated lines,totem patterns and color symbolized nobility,leath,marble and color tone enrich the sentiment of space.Exqusite technique and layering expression make space a splendid European palace,showing dignity of noble. Different colors are added in different space.Complicated lines and tensional color tone are simplified,classical elements are reserved. The sense of romance was setted off in light color, a tranquil and elegant feeling was showed in the whole space.

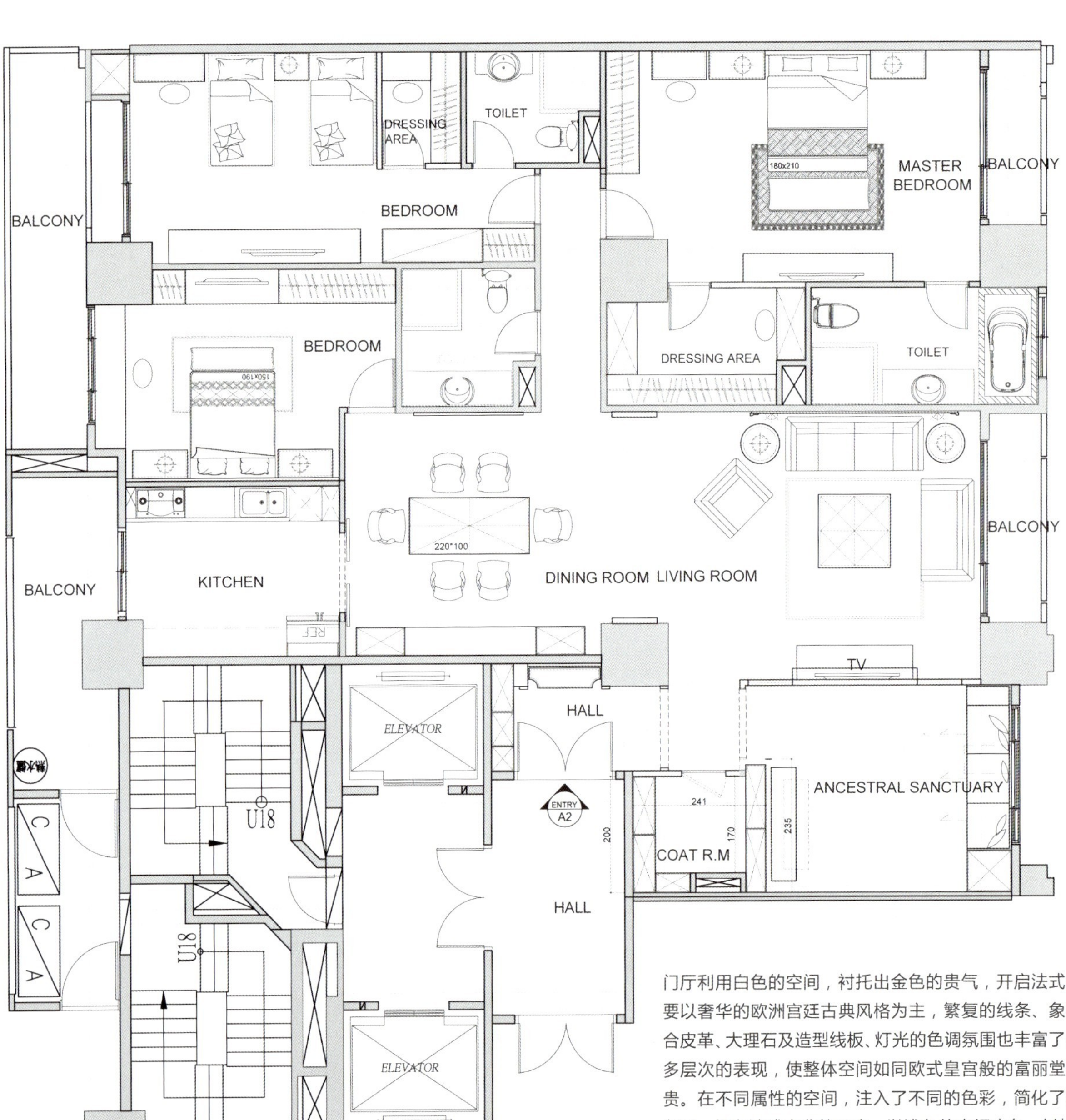

Plan 平面图

门厅利用白色的空间，衬托出金色的贵气，开启法式古典的尊贵序幕。空间主要以奢华的欧洲宫廷古典风格为主，繁复的线条、象征贵气的图腾及色彩，结合皮革、大理石及造型线板、灯光的色调氛围也丰富了空间情调。以细腻的手法多层次的表现，使整体空间如同欧式皇宫般的富丽堂皇，呈现出宫廷贵族的尊贵。在不同属性的空间，注入了不同的色彩，简化了繁复的线条及充满张力的色调，保留法式古典的元素，以浅色的空间底色，衬托出整体空间的浪漫质感呈现静谧而优雅的贵族气息。

ZHONG YUE ZHI BAO

中悦至宝

Design Company: Heyang Design
设计公司：禾洋设计

Designer: Xie Qiugui
设计师：谢秋贵

Location: Taiwan
项目地点：台湾

Area: 231m²
项目面积: 231 平方米

Marble floor and ceiling are exquisite and are full of three-dimensional,layered performance. The wall back of sofa is electric carved with totem to match the overall style,gorgeous and tasteful.The TV wall is modeled like a neoclassical fireplace,consisting an elegant,stately posture.The Spanish aurora marble with special shape attracts people's attention.Dining room contains a 8-persons table,chandelier is installed in a round ceiling to build a romantic warmth atmosphere. Sunlight goes through the French window,a big antique vase matches light,reflecting a dignity and graceful mien.

Custom furnitures in master room reflect an extraordinary taste in both modeling and color.The bed in girl's room is exquisitely handmade with famous diamond,reflecting a tender dignity,satisfying girl's romantic dreams. Culture forms a harmony with architect features perfectly,diverse exquisite elements are applied,like a magnificent movement of an immortal epic,creating a extradinary luxury house as owner's tailor.

客厅地坪大理石选择及顶棚造型富有层次和立体感，极其讲究，在沙发背墙选用电射图腾雕刻以银色与家具和整体风格作搭配，华丽却不失品位，在电视主墙以新古典意象的壁炉造型为基础，架构出优雅华丽的大器姿态，西班牙奥罗拉大理石抢眼的外形立刻吸引众人的目光。移步来到餐厅场域，敞阔的空间可容纳下长形八人座的餐桌椅，为了餐叙氛围温馨浪漫，在顶棚以圆形上潜造型搭配水晶吊灯铺陈；落地窗外洒入明亮的采光，一方大型古董珍藏花器及灯光规划，整体呈现出雍容华贵的迷人风采。

主卧配置订制家具，不论是外形还是色泽皆流露顶级不凡的品位。女孩房床绷皮以精致且繁复手工制作而成，镶上知名水钻更显娇嫩贵气，满足少女梦幻的浪漫情怀。 完美结合人文与丰富的建筑特色，灵活运用各种细腻精致的元素，如同不朽史诗中的磅礴乐章，为使用者量身打造隽永非凡的奢华大宅！

The first floor plan 一楼平面图

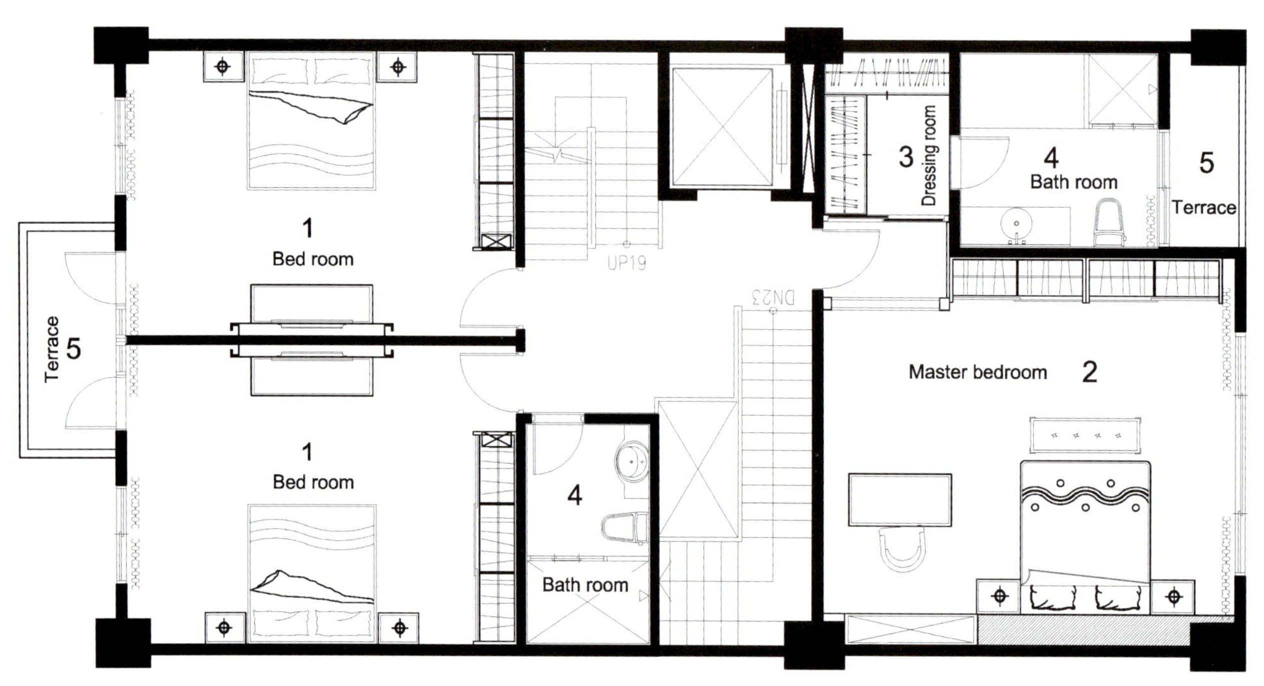

The second floor plan 二楼平面图

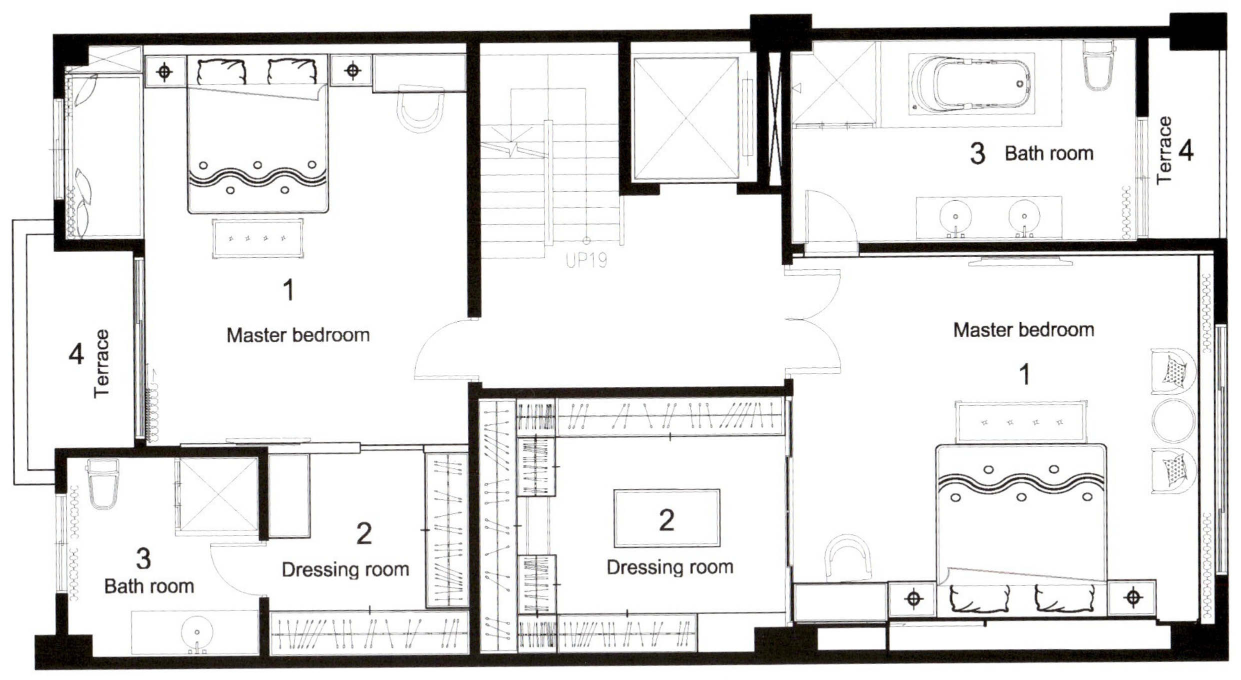

The third floor plan 三楼平面图

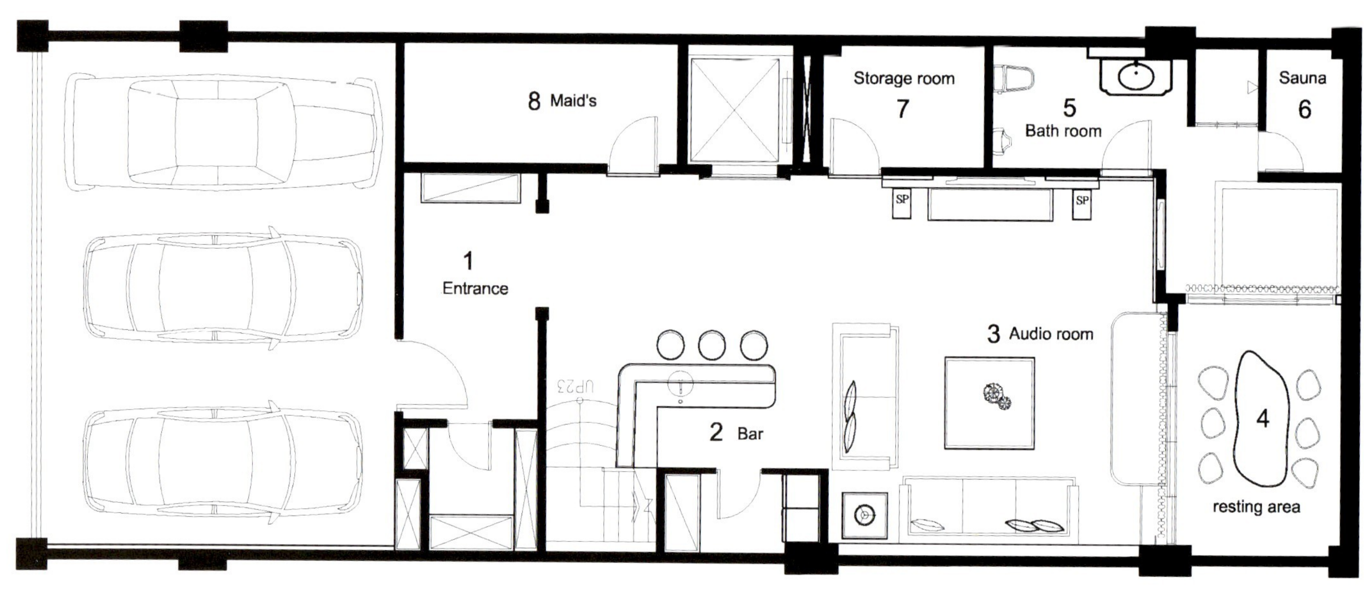

The basement plan 地下室平面图

图书在版编目（CIP）数据

古典奢华／深圳市海阅通文化传播有限公司主编.
北京：中国建筑工业出版社，2013.4
（居家空间创意集）
ISBN 978-7-112-15184-4

Ⅰ.①古··· Ⅱ.①深··· Ⅲ.①住宅—室内装饰设计—图集 Ⅳ.
①TU238

中国版本图书馆CIP数据核字(2013)第038819号

责任编辑：费海玲　张幼平　王雁宾
责任校对：姜小莲　陈晶晶
装帧设计：王　硕
采　　编：李箫悦　罗　芳

居家空间创意集
古典奢华
深圳市海阅通文化传播有限公司　主编
*
中国建筑工业出版社出版、发行（北京西郊百万庄）
各地新华书店、建筑书店经销
深圳市海阅通文化传播有限公司制版
北京方嘉彩色印刷有限责任公司印刷
*
开本：880×1230毫米　1/16　印张：$5^1/_2$　字数：180千字
2013年 6月第一版　2013年 6月第一次印刷
定价：29.00元
ISBN 978-7-112-15184-4
(23278)